Interactive
Mathematics Program®

INTEGRATED HIGH SCHOOL MATHEMATICS

The Diver Returns

FIRST EDITION AUTHORS:
Dan Fendel, Diane Resek, Lynne Alper, and Sherry Fraser

CONTRIBUTORS TO THE SECOND EDITION:
Sherry Fraser, IMP for the 21st Century
Jean Klanica, IMP for the 21st Century
Brian Lawler, California State University San Marcos
Eric Robinson, Ithaca College, NY
Lew Romagnano, Metropolitan State College of Denver, CO
Rick Marks, Sonoma State University, CA
Dan Brutlag, Meaningful Mathematics
Alan Olds, Colorado Writing Project
Mike Bryant, Santa Maria High School, CA
Jeri P. Philbrick, Oxnard High School, CA
Lori Green, Lincoln High School, CA
Matt Bremer, Berkeley High School, CA
Margaret DeArmond, Kern High School District, CA

Key Curriculum Press

Second Edition   I M P

This material is based upon work supported by the National Science Foundation under award numbers ESI-9255262, ESI-0137805, and ESI-0627821. Any opinions, findings, and conclusions or recommendations expressed in this publication are those of the authors and do not necessarily reflect the views of the National Science Foundation.

Key Curriculum Press
1150 65th Street
Emeryville, California 94608
email: editorial@keypress.com
www.keypress.com
10 9 8 7 6 5 4 3 2 1   15 14 13 12 11
ISBN 978-1-60440-144-8
Printed in the United States
of America

Project Editors
Mali Apple, Josephine Noah

Project Administrator
Emily Reed

Professional Reviewers
Rick Marks, Sonoma State University, CA
D. Michael Bryant, Santa Maria High School, CA, retired

Accuracy Checker
Carrie Gongaware

First Edition Teacher Reviewers
Kathy Anderson, Aptos High School, CA
Dan H. Brutlag, Tamalpais High School, CA
Robert E. Callis, Hueneme High School, CA
Susan Schreibman Ford, Delhi High School, CA
Mary L. Hogan, Arlington High School, MA
Jane M. Kostik, Patrick Henry High School, MN
Brian Lawler, California State University San Marcos, CA
Brent McClain, Vernonia School District, OR
Michelle Novotny, Eaglecrest High School, CO
Barbara Schallau, East Side Union High School District, CA
James Short, Oxnard Union High School District, CA
Kathleen H. Spivack, Wilbur Cross High School, CT
Linda Steiner, Orange Glen High School, CA
Marsha Vihon, Corliss High School, IL
Edward F. Wolff, Arcadia University, PA

First Edition Multicultural Reviewers
Genevieve Lau, Ph.D., Skyline College, CA
Luís Ortiz-Franco, Ph.D., Chapman University, CA
Marilyn Strutchens, Ph.D., Auburn University, AL

Copyeditor
Brandy Vickers

Interior Designer
Marilyn Perry

Production Editor
Andrew Jones

Production Director
Christine Osborne

Editorial Production Supervisor
Kristin Ferraioli

Compositors
Kristin Ferraioli, Maya Melenchuk

Art Editor/Photo Researcher
Maya Melenchuk

Technical Artists
Lineworks, Inc., Maya Melenchuk, Kristin Ferraioli

Illustrator
Juan Alvarez, Alan Dubinsky, Tom Fowler, Nikki Middendorf, Briana Miller, Evangelia Philippidis, Paul Rodgers, Sara Swan, Martha Weston, April Goodman Willy, Amy Young

Cover Designer
Jenny Herce

Printer
Lightning Source, Inc.

Executive Editor
Josephine Noah

Publisher
Steven Rasmussen

CONTENTS

The Diver Returns—Circular
Functions, Vector Components,
and Complex Numbers

The Diver Returns

Circular Functions, Vector Components,
and Complex Numbers

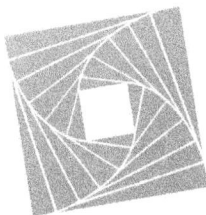

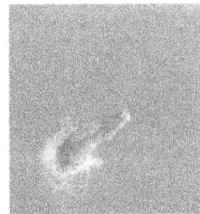

The Diver Returns—Circular Functions, Vector Components, and Complex Numbers

Back to the Circus

In the Year 3 unit *High Dive*, you analyzed a circus act in which a diver falls from a turning Ferris wheel into a tub of water carried by a moving cart.

That version of the problem involved a major simplification. Now you get to examine a more realistic—but more complex— version of the problem, which will lead you to several important new mathematical ideas.

To begin, you will review the essential elements that go into solving the original problem.

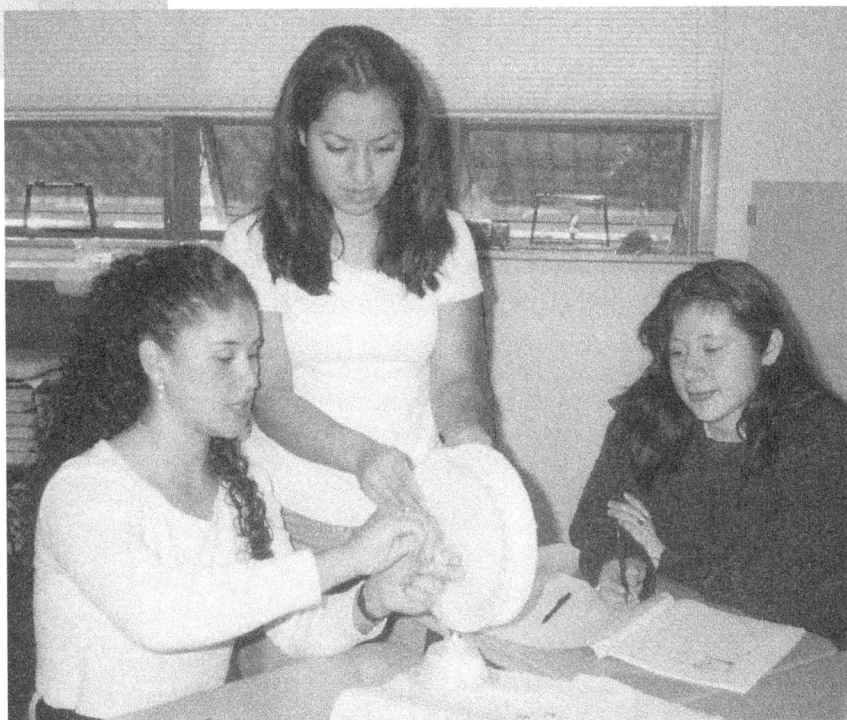

Maribel DeLoa, Vivian Barajas, and Caroline Moo build a physical model as a first step toward solving the original unit problem.

The Circus Act

In the circus act from *High Dive,* a diver's platform is attached to one of the seats on a Ferris wheel. The platform sticks out, perpendicular to the plane of the Ferris wheel. A tub of water is on a moving cart that runs along a track, in the plane of the Ferris wheel, and passes under the end of the platform.

As the Ferris wheel turns, an assistant holds the diver by the ankles. The assistant must let go at exactly the right moment, so that the diver will land in the moving tub of water.

In the *High Dive* analysis of this situation, you made the simplifying assumption that when the assistant lets go, the diver falls *as if he were being dropped from a stationary platform.*

continued ▶

In this unit, *The Diver Returns,* you will eliminate that simplifying assumption. The Ferris wheel is actually moving constantly. Its motion means that the diver is also moving, even before he is dropped. For instance, if his platform is moving upward at the moment he is let go, the diver actually starts to go up before going down. And if the platform is moving sideways to some extent as the Ferris wheel turns, the diver moves in that direction too.

You will deal with these complexities in this unit. In other words, the unit problem is to investigate this question:

> *Taking into account the diver's initial motion due to the movement of the platform, when should the assistant let go?*

In this activity, you will rely on your intuition. Later in the unit, you will examine in detail the mathematics and physics of the situation.

Imagine that you are holding on at a point on the circumference of the Ferris wheel. The Ferris wheel is turning rapidly counterclockwise. Suddenly, you let go.

Sketch the path of your motion until you reach the ground. Do this for several positions around the wheel.

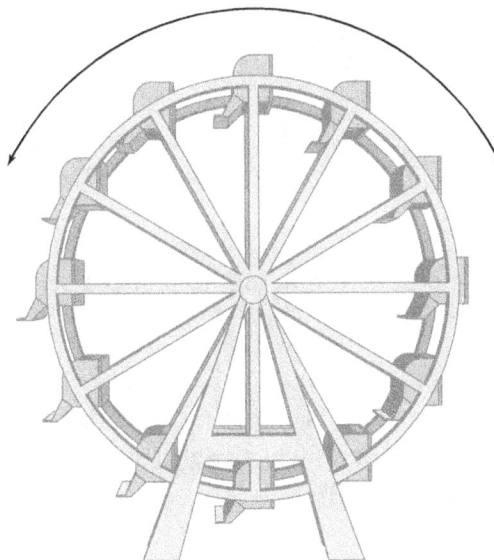

As the Ferris Wheel Turns

Before examining the details of the more complex Ferris wheel problem, you will reexamine the original problem. Here is some basic information:

- The radius of the Ferris wheel is 50 feet.
- The Ferris wheel turns counterclockwise at a constant speed and makes a complete turn every 40 seconds.
- The center of the Ferris wheel is 65 feet off the ground.

1. At what speed is the platform moving (in feet per second) as it goes around on the Ferris wheel?

2. The rate at which an object turns is called *angular speed*, because it measures how fast an angle is changing. Angular speed does not depend on the radius. Through what angle (in degrees) does the Ferris wheel turn each second?

continued ▶

You can describe positions in the cycle of the Ferris wheel in terms of the face of a clock. For example, the highest point in the wheel's cycle is the 12 o'clock position. The point farthest to the right is the 3 o'clock position.

3. How many seconds does it take the platform to go each of these distances?

   a. From the 3 o'clock to the 11 o'clock position

   b. From the 3 o'clock to the 7 o'clock position

   c. From the 3 o'clock to the 4 o'clock position

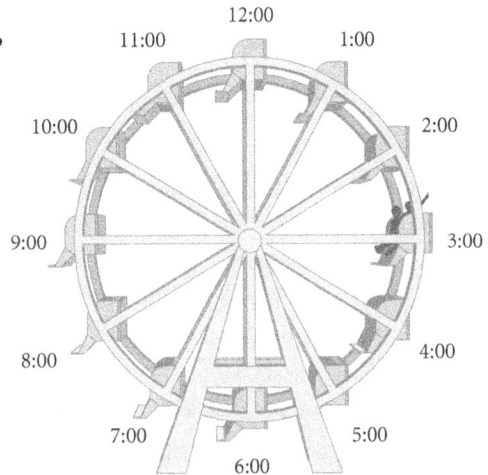

4. What is the platform's height off the ground at each of these times?

   a. 1 second after passing the 3 o'clock position

   b. 6 seconds after passing the 3 o'clock position

   c. 14 seconds after passing the 3 o'clock position

   d. 23 seconds after passing the 3 o'clock position

Graphing the Ferris Wheel

In Question 1, use these basic facts about the Ferris wheel:

- The Ferris wheel has a radius of 50 feet.
- The center of the Ferris wheel is 65 feet off the ground.
- The Ferris wheel turns counterclockwise at a constant speed, making a complete turn every 40 seconds.
- At the starting time, $t = 0$, the diver's platform is at the 3 o'clock position in its cycle.

1. Draw a graph showing the platform's height off the ground, h, as a function of t. Show two full cycles of the Ferris wheel's movement on your graph.

2. Draw a graph showing the platform's horizontal coordinate, x (relative to the center of the Ferris wheel), as a function of t. Again, show two full cycles.

3. Describe how your two graphs would change if you made each of the adjustments in parts a to c. Treat each part separately, changing only the item mentioned.

   a. How would the graphs change if the radius of the Ferris wheel were smaller?

   b. How would the graphs change if the Ferris wheel were turning faster—that is, if the period were shorter?

   c. How would the graphs change if you measured height with respect to the center of the Ferris wheel instead of with respect to the ground? For example, if the platform were 40 feet above the ground, you would treat this as a height of -25 feet, because 40 feet above the ground is 25 feet below the center of the wheel.

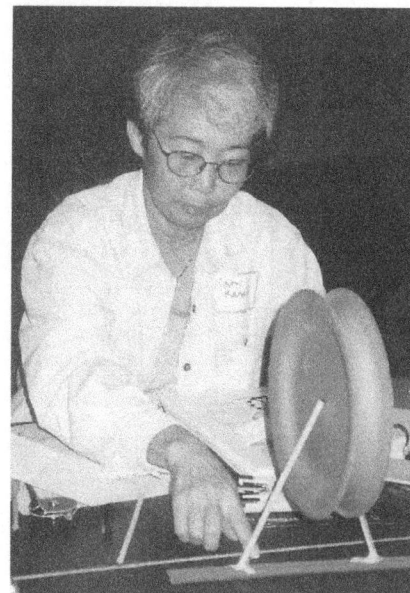

Distance with Changing Speed

1. Curt is traveling home from college to visit his family. He drives from 1 p.m. to 3 p.m. at an average speed of 50 miles per hour. Then he drives from 3 p.m. to 6 p.m. at an average speed of 60 miles per hour.

   a. Draw a graph showing Curt's speed as a function of time for the entire period from 1 p.m. to 6 p.m. Treat his speed as constant for each of the two time periods—from 1 p.m. to 3 p.m. and from 3 p.m. to 6 p.m.

   b. Describe how to use areas in your graph to represent the total distance Curt travels.

2. A triathlete is running at a steady speed of 20 feet per second. At exactly noon, she starts to increase her speed. Her speed increases at a constant rate so that 20 seconds later, she is going 30 feet per second.

   a. Graph the runner's speed as a function of time for the 20-second interval beginning at noon.

   b. Calculate the runner's average speed for this 20-second interval.

   c. Explain how to use area to find the total distance she runs during this 20-second interval.

Free Fall

Using experiments and theoretical analysis, physicists have confirmed this principle about falling objects:

Falling objects have constant acceleration.

This principle assumes there is no air resistance or other complicating factors to interfere with an object's fall. That is, the principle describes the behavior of *free-falling* objects. In this unit, unless you are told otherwise, assume that falling objects are falling freely.

This broad principle of free-falling objects can be stated more precisely:

The instantaneous speed of a free-falling object increases approximately 32 feet per second for each second of the object's fall.

The simplest case is when an object starts from rest—that is, when its speed is zero at $t = 0$. The object's instantaneous speed after 1 second is 32 feet per second; after 2 seconds, its instantaneous speed is 64 feet per second; and so on.

continued ▶

Your task is to use these principles to express the distance an object falls in terms of the amount of time it has been falling. Assume that the object is dropped from rest and falls freely.

1. a. How fast is the object going at $t = 5$?

   b. How far does the object fall in the first 5 seconds?

2. Generalize your work from Question 1 to develop an algebraic expression for how far the object falls in the first t seconds.

3. Suppose the object starts falling from a height of h feet. What is its height after t seconds? Assume the object has not yet reached the ground.

4. Use your result from Question 3 to find an expression in terms of h for the time it would take the object to reach the ground.

Now apply your work to a simple version of the circus act.

5. Suppose the platform is fixed at 90 feet above the ground, the diver falls freely from rest, and the water level in the tub is 8 feet above the ground. How long will it take the diver to reach the water?

Moving Cart, Turning Ferris Wheel

You will now re-create the solution to the simplified version of the circus act. Recall these facts about the moving cart containing the tub of water:

- When the cart starts moving, it is 240 feet to the left of the center of the Ferris wheel's base.
- The cart travels to the right at a constant speed of 15 feet per second.
- The water level in the cart is 8 feet above the ground.
- When the cart starts moving, the diver's platform is at the 3 o'clock position in its cycle.

This simplified version assumes that at the moment of release, the diver falls freely from rest.

Your task is to figure out when the assistant should let go of the diver. Let $t = 0$ represent the moment the platform passes the 3 o'clock position. Let W represent the number of seconds until the release of the diver. You need to determine the right value for W.

In addition to giving the value of W, also determine these things:

- Where the platform will be in the Ferris wheel's cycle when the diver is dropped
- Where the cart will be when the diver hits the water

Which Weights Weigh What?

Do you remember the economical king with the different kinds of scales? You first met him in the Year 1 POWs *Eight Bags of Gold* and *Twelve Bags of Gold,* and he reappeared in the Year 3 POW *And a Fortune, Too!* He has now decided that an old-fashioned balance scale is what he wants. But he wants to be able to use it as efficiently as possible.

In the king's country, it is customary to pay for food by weight, but the king doesn't trust the merchants' scales. He insists on using his balance scale to verify their claims.

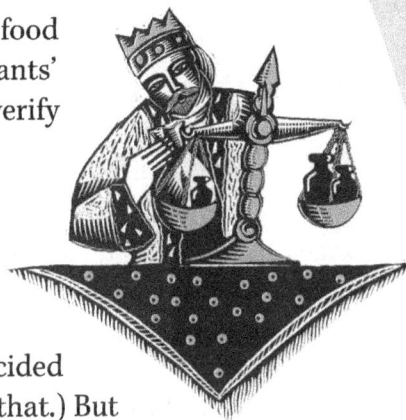

He has decided that he needs only to verify the weights of packages that weigh whole numbers of ounces. He wants to be able to verify weights starting from 1 ounce and including all possibilities up to some still-undecided value. (He's the king, so he can make rules like that.) But he's concerned that in order to weigh different amounts, he might need a large set of standard weights—a 1-ounce weight, a 2-ounce weight, a 3-ounce weight, and so on.

o *Method 1: Combining Weights on One Side*

One of the king's advisers points out that the situation is not so bad. For example, the king doesn't need a 3-ounce weight. He can verify the weight of a 3-ounce package by balancing it against a combination of a 1-ounce and a 2-ounce weight, as shown here.

continued

The king is relieved to hear this, but he still needs to decide what weights to buy. No set of weights will cover all possibilities, so the choice seems to depend on how many weights he is willing to buy.

Of course, if he buys only one weight, it must be a 1-ounce weight. Unfortunately, that would allow him to verify the weight of only a 1-ounce package.

Suppose the king is willing to buy two weights. Exactly which weights should he buy in order to verify the weight of as heavy a package as possible (and still include all whole-number possibilities of lesser weights)? And what if he buys three weights? Or four?

Method 2: Using Weights on Both Sides

Another adviser suggests that perhaps the king can omit both the 2-ounce weight and the 3-ounce weight. For example, to verify a merchant's claim that a package weighs 3 ounces, the king can put the package together with a 1-ounce weight on one side of the balance scale and put a 4-ounce weight on the other side, as shown here. If the two sides balance, the king will know that the package really does weigh 3 ounces. But he will need something other than the 1-ounce and 4-ounce weights to check a 2-ounce package.

This method seems to provide more options, meaning the king may be able to weigh more possibilities for a given number of weights. As with the first method, if the king buys only one weight, it will need to be a 1-ounce weight. But that will allow him to verify the weight of only a 1-ounce package.

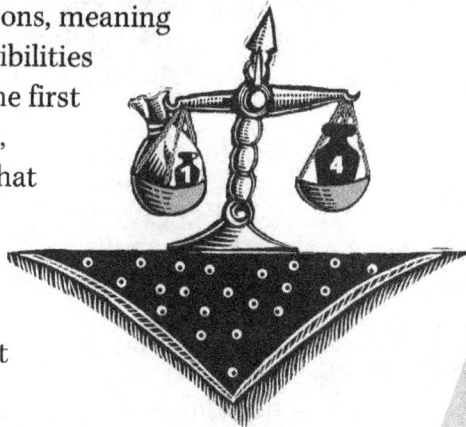

Suppose the king is willing to buy two weights. Which should he buy? And what if he buys three weights? Or four?

continued

○ *Your Task: Consider Each Method*

First consider Method 1, in which the weights all go on one side of the balance scale and the package goes on the other side. Start with the specific cases of two weights, three weights, and four weights. Tell the king which weights he should choose in each case and what package weights his choice will allow him to verify. Then look for a general procedure that tells him which weights to choose and what package weights he will be able to verify.

Next, consider Method 2, which allows the king to put weights on both sides of the balance scale. With this method, he can still put the weights all on one side if he chooses. Again, start with specific cases of two weights, three weights, and four weights. Tell the king which weights he should choose in each case and what package weights his choice will allow him to verify. Then look for a general procedure that tells him which weights to choose and what package weights he will be able to verify.

○ *Write-up*

Your write-up should contain these components:

1. *Problem Statement*

2. *Process*

3. *Solution:* For each method, include a clear explanation of why the weights you choose are the best. State how high your choice of weights will allow the king to go, and explain this result. Include a general solution if you can.

4. *Self-assessment*

The Standard POW Write-up

Each Problem of the Week is unique, so the form of the write-up may vary from one POW to the next. Nevertheless, most of the categories you will use for your POW write-ups will be the same throughout the year. The list below summarizes the standard categories.

Some POW write-ups will use other categories or require more specific information within a particular category. If the write-up instructions for a given POW simply list a category by name, however, use these descriptions.

1. *Problem Statement:* State the problem clearly and in your own words. Your problem statement should be clear enough that someone unfamiliar with the problem could understand what you are being asked to do.

2. *Process:* Describe what you did in attempting to solve the problem. Use your notes as a reminder. Include things that didn't work out or that seemed like a waste of time. Do this part of the write-up even if you didn't solve the problem.

    If you get assistance of any kind on the problem, tell what the assistance was and how it helped you.

3. *Solution:* State your solution as clearly as you can. Explain how you know that your solution is correct and complete. If you obtained only a partial solution, give that. If you were able to generalize the problem, include your general results.

    Write your explanation in a way that will be convincing to someone else—even someone who initially disagrees with your answer.

4. *Self-assessment:* Tell what you learned from this problem. Be as specific as you can. Also assign yourself a grade for your work on this POW, and explain why you think you deserve that grade.

A Simple Summary and a Complex Beginning

Part I: A Simple Summary

You have now reexamined the simplified version of the circus act problem.

1. Put together a summary of the formulas you used in solving this problem.

2. Describe briefly how you used these formulas to solve the problem.

Part II: A Complex Beginning

You are about to start work on the more complex version of the Ferris wheel problem. To get started thinking about how the Ferris wheel's motion affects the diver's fall, consider a situation that involves circular motion but not gravity.

3. You are a skateboarder. You go to a skateboard park that has a merry-go-round. You hold on to the railing while the merry-go-round spins rapidly, so you are moving in a circular path while on your skateboard.

Suddenly, you let go of the railing. What is the path of your motion? Assume your skateboard has perfect ball bearings and that the merry-go-round has an ideal surface so that there is no friction to slow you down.

A Falling Start

In *Moving Cart, Turning Ferris Wheel,* you analyzed the Ferris wheel problem as if the diver were to fall from rest—that is, as if the wheel weren't moving. But the diver is moving while he's still on the Ferris wheel, because the platform itself is moving. This means he will have some initial speed at the moment he is released.

The simplest cases to consider are those in which the diver's initial motion is vertical—either straight up or straight down. Analyzing these cases will require you to take a new look at the formulas related to falling objects, as well as consider some new ideas about solving equations.

Zoë O'Rorke tries to visualize what effect the motion of the Ferris wheel will have on the diver's path upon release.

Look Out Below!

Maxine lives in a high-rise dormitory on her college campus. One afternoon, while studying in her room, she hears a voice shout, "Hey, up there! Could you toss me my pillow?"

She glances out the floor-to-ceiling window just in time to see the pillow fly past.

In this activity, you will analyze the pillow's fall. Because pillows are comparatively light for their size, you can't ignore the effect of air resistance on a falling pillow. Assume that the falling pillow accelerates at a rate of only 20 feet per second for each second it falls (although this is an oversimplification). Also assume that the pillow is traveling at an instantaneous speed of 30 feet per second when Maxine sees it.

1. What is the pillow's *instantaneous* speed 1 second after Maxine sees it? Two seconds after she sees it?

2. What is the pillow's *average* speed for the first 2 seconds after Maxine sees it?

3. How far does the pillow fall during the first 2 seconds after Maxine sees it?

Maxine looks down at the sidewalk and sees someone pick up the pillow. The sidewalk is 200 feet below her window.

4. How long does it take the pillow to reach the ground from the time Maxine sees it? Give your answer to the nearest tenth of a second.

5. Find a general expression for the height of the pillow t seconds after Maxine sees it.

The Diver and the POW

Part I: What Happens to the Diver?

You have seen something about what happens to objects when they start out moving in a circular path and are then released. Now you will look at how these principles affect our circus diver.

In thinking about these questions, pay special attention to the cases in which the diver is released from the 3 o'clock, 6 o'clock, 9 o'clock, and 12 o'clock positions.

1. From which release positions will the diver's falling time be increased due to his initial motion from the Ferris wheel? From which release positions will the falling time be decreased?

2. From which release positions will the diver land to the left of his release position? From which release positions will he land to the right?

3. In *Moving Cart, Turning Ferris Wheel,* using the assumption that the diver falls straight down as if dropped from rest, you found that the assistant should hold on to the diver for about 12.3 seconds. This means the diver will be released between the 12 o'clock and 11 o'clock positions.

   Take into account the effect of the Ferris wheel's motion on the diver's fall. Do you think the assistant should hold on to the diver longer than 12.3 seconds or let go sooner than that? Explain your reasoning.

Part II: A POW Beginning

This problem is related to POW 1: *Which Weights Weigh What?*

4. Suppose you are camping and have a pancake recipe that calls for exactly 10 ounces of milk. Unfortunately, you brought only 8-ounce and 6-ounce cups. How can you use these two cups to measure exactly 10 ounces? What other amounts could you measure exactly?

5. Discuss how Question 4 is mathematically similar to the POW and how it is different.

Big Push

You have solved the Ferris wheel problem based on the simplifying assumption that when the diver is released, he falls as if the Ferris wheel has not been moving. Now you will examine, in a specific case, how the Ferris wheel's motion affects the diver's fall.

Here is a summary of some important facts about the situation:

- The Ferris wheel has a radius of 50 feet.
- The center of the Ferris wheel is 65 feet above the ground.
- The Ferris wheel turns counterclockwise at a constant rate, making a complete turn every 40 seconds.
- The water level in the cart is 8 feet above the ground.

Imagine that the assistant lets go of the diver at the 9 o'clock position. Because the platform is moving downward at that moment, the diver is also moving downward as he is released. His initial speed when released is equal to the speed with which he was moving when he was on the platform. You have already found this to be 2.5π feet per second, or approximately 7.85 feet per second.

continued ▶

After the diver is released, his speed will increase. As with any object falling freely, the diver's speed increases by 32 feet per second for each second he falls. For now, assume the tub of water is in a fixed position, directly below the point of release.

1. What is the diver's height t seconds after he is released?

2. How long will it take from the time the diver is released until he reaches the water?

3. Suppose the Ferris wheel is not moving and the diver is simply dropped, with no initial speed, from the 9 o'clock position. How long will it take the diver to fall to the water level?

Now recall the fact that the cart actually travels to the right at a constant speed of 15 feet per second.

4. Compare your answers to Questions 2 and 3. Determine how far the cart would travel during a time interval equal to the difference between those answers.

Finding with the Formula

Sometimes, when repeating a mathematical process over and over, it becomes easier to develop a formula that gives you the same result.

Solving quadratic equations is one of those situations. You can use the method of completing the square for individual examples, but there is a general formula that saves the trouble of repeating the steps each time.

The general **quadratic equation** is usually written in the form $ax^2 + bx + c = 0$. The coefficients a, b, and c can be any numbers, except that a cannot be 0. If you apply the method of **completing the square** to the general quadratic equation, you get an expression, called the **quadratic formula,** that gives the solutions in terms of a, b, and c. The general result says

$$\text{If } ax^2 + bx + c = 0, \text{ and } a \neq 0, \text{ then } x = \frac{-b \pm \sqrt{b^2 - 4ac}}{2a}.$$

That is, if x is a solution to the equation $ax^2 + bx + c = 0$, then x must be equal to either

$$\frac{-b + \sqrt{b^2 - 4ac}}{2a} \quad \text{or} \quad \frac{-b - \sqrt{b^2 - 4ac}}{2a}$$

1. Use the quadratic formula to solve the equation $x^2 - 3x - 28 = 0$. Check your answers.

2. Use the quadratic formula to solve the equation $3x^2 + 7x = 5$. Check your answers.

Using Your ABCs

1. Find all the solutions to each of these equations using the quadratic formula. Give exact solutions, using square roots if necessary. Then approximate the solutions to the nearest tenth, and use the equations to confirm that these solutions seem correct.

   a. $x^2 + 7x + 12 = 0$

   b. $x^2 - 3x - 8 = 0$

   c. $2x^2 + 5x - 1 = 0$

   d. $2x^2 - 3x + 4 = 0$

2. Set up and solve a quadratic equation to answer each of these questions.

   a. A rectangle with one side 5 feet longer than the other side has an area of 126 square feet. What are the rectangle's dimensions?

   b. A right triangle has one leg 6 inches shorter than the other leg, and its hypotenuse is 13 inches long. How long are the triangle's legs?

   c. An object is thrown up into the air from the roof of a building. Its height above the ground h (in feet) after t seconds is given by the equation $h = 90 + 50t - 16t^2$. When will the object be 120 feet high?

Imagine a Solution

Solving quadratic equations is closely related to finding the x-intercepts of graphs of **quadratic functions**. Because some of these graphs don't have any x-intercepts, it makes sense that some quadratic equations won't have any solutions.

The equation $x^2 + 1 = 0$ is an example of a quadratic equation with no solution (and the function $y = x^2 + 1$ has no x-intercepts). You can rewrite this equation as the equivalent equation $x^2 = -1$ to see that neither a positive nor a negative number can be a solution.

Suppose, though, that we make up a new number that would be the solution to this equation. This new number might help us find solutions to many other equations as well.

Several centuries ago, some mathematicians decided to see what would happen if they did this. The number they invented is now represented by the symbol i, so $i^2 = -1$ and $\sqrt{-1} = i$. This led to a whole new family of numbers, such as $2i$, $-i$, and so on. These are called **imaginary numbers.** (Numbers that appear on the number line, such as 3, -7, 0, 6.98, and $\sqrt{2}$, are called **real numbers.**)

continued ▶

1. Write each square root in terms of this new number i. Explain your reasoning.

   a. $\sqrt{-25}$

   b. $\sqrt{-100}$

2. Use your reasoning from Question 1 to find solutions to each equation.

   a. $x^2 + 9 = 0$

   b. $x^2 + 62 = 13$

   c. $x^2 - 10 = -15$

3. a. Investigate what happens when you raise i to different powers. For example, find the value of i^3, i^4, i^5, and so on.

   b. Use your results to find a simpler form of i^{3057}. Explain your answer.

   c. Write a general procedure for finding the value of i^n in simple form without doing lots of repetition.

Complex Numbers and Quadratic Equations

The imaginary numbers i and $-i$ are the solutions to the quadratic equation $x^2 + 1 = 0$. Similarly, $2i$ and $-2i$ are the solutions to the equation $x^2 + 4 = 0$. There are even imaginary numbers like $\sqrt{2} \cdot i$, which is a solution to the equation $x^2 + 2 = 0$.

Imaginary numbers provide solutions to many quadratic equations that have no real-number solutions. But some quadratic equations need a combination of a real and an imaginary number. Such combinations are called **complex numbers.** Each complex number can be expressed as the sum of a real number and an imaginary number, such as $2 + 3i$ and $-7 - 2i$ (which is equivalent to the sum $-7 + -2i$). Complex numbers turn up as solutions to many quadratic equations.

The general form of a complex number is $a + bi$. The term a is called the *real part*. The term bi is called the *imaginary part*. Either part of a complex number could be zero. This activity is about the simplest complex number that has both a nonzero real part and a nonzero imaginary part: $1 + i$.

1. Show that $1 + i$ is a solution to the equation $x^2 - 2x + 2 = 0$.

2. Apply the quadratic formula to the equation $x^2 - 2x + 2 = 0$ to find a second solution.

3. Verify that the second solution satisfies the equation.

4. Use the quadratic formula to find the solutions to the equation $x^2 - 4x + 7 = 0$. Check both solutions.

Complex Components

In the unit problem, at the moment of release, the diver is most likely moving sideways as well as up or down. That is, his **velocity** has two components—vertical and horizontal. The complex numbers you encountered in working with quadratic equations also have two components—real and imaginary. Interestingly, you can represent these two number components using vertical and horizontal axes, with good results.

Consider the complex number $2 + 3i$, with a real part of 2 and an imaginary part of $3i$. If you treat the values 2 and 3 as x- and y-coordinates, the number $2 + 3i$ corresponds to the point $(2, 3)$ in the coordinate plane. Notice that the axes are labeled "Real axis" and "Imaginary axis."

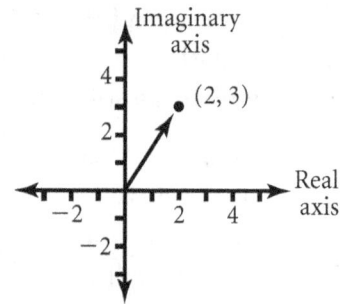

In this context, the coordinate system is called the **complex plane.** We often represent points in the complex plane by using arrows from the origin, as shown here. These arrows are called **vectors.**

You can think of a "trip" from the origin to $(2, 3)$ as made up of two parts—two steps to the right and three steps up. This movement is shown by the gray arrows in the second diagram. This reflects the idea that the vector for $2 + 3i$ is the "sum" of a horizontal vector for 2 and a vertical vector for $3i$.

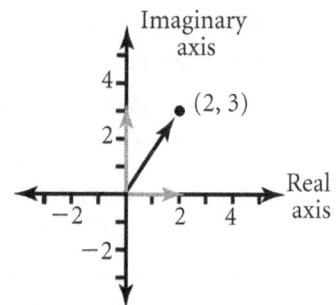

1. Construct a complex plane and draw a single vector representing the complex number $3 + i$. In the same diagram, draw a vector representing $1 + 4i$.

2. Find the sum of the two complex numbers from Question 1. Represent the sum as a vector on your diagram.

3. Examine the shape formed by the origin, the two complex numbers $3 + i$ and $1 + 4i$, and their sum. What do you notice? How is the vector for the sum related geometrically to the vectors for $3 + i$ and $1 + 4i$?

Three O'Clock Drop

In *Big Push*, you found the falling time for the diver if he is released from the 9 o'clock position. In that problem, the motion of the Ferris wheel gives the diver an initial downward velocity as he is released. You saw that it will take him less time to fall than if he is dropped from a stationary Ferris wheel.

Now consider what happens if the diver is released from the 3 o'clock position. Because the platform is moving upward at that moment, he will start off with an initial upward motion. As in *Big Push*, his initial speed is equal to the speed with which he was moving when he was on the platform.

Assume again that the cart is in a fixed position, directly below the diver's point of release.

1. How long will it take from the time the diver is released until he reaches the water?

2. How long would it have taken the diver to reach the water if he had been released from a *motionless* Ferris wheel (from the 3 o'clock position)?

Up, Down, Splat!

Melissa's science class is having a contest. The challenge is to build a container that will keep an egg from breaking when dropped from a particular window.

Melissa is quite confident of her contraption. She leans out the window, which is 25 feet off the ground, and hurls her egg container straight up in the air with an initial velocity of 35 feet per second. (Consider velocity upward to be positive.)

Assume the egg container's velocity is affected by gravity in the usual way. That is, the velocity decreases by 32 feet per second for each second the container travels.

1. How long does it take the container to hit the ground?

2. At what speed does the container hit the ground?

Falling Time for Vertical Motion

In each of several recent problems, you figured out how long it would take an object to fall a certain distance if it started with a certain initial velocity.

When an object is falling freely, its height after t seconds is given by the expression $h + vt - 16t^2$. In this expression, h is the object's initial height and v is the object's initial velocity upward or downward (where upward motion is considered positive). Finding out when the object will hit the ground is equivalent to solving the equation $h + vt - 16t^2 = 0$.

In specific cases, you might be able to solve the equation (or get a good estimate) by guess-and-check or with a graph. But when you solve the main unit problem, you will not have numerical values for h or v, because those coefficients will be expressed in terms of the variable W.

In preparation for dealing with that complication, your task here is to solve the equation $h + vt - 16t^2 = 0$ in terms of h and v.

1. Rewrite the equation $h + vt - 16t^2 = 0$ in the standard form $ax^2 + bx + c = 0$, with a positive value for a.

2. Use the quadratic formula to solve your equation from Question 1. Your answer should give t in terms of h and v.

3. Which of the two solutions you found in Question 2 will give you a positive value for t? Assume that h is a positive number.

Isaac Newton (1643–1727), an English mathematician and physicist, is one of the most influential scientists of all time. The laws of motion and gravity that he described are still studied by students around the world.

Components of Velocity

In several recent activities, you've considered objects with either positive or negative initial vertical velocity. But when the assistant releases the Ferris wheel diver, he might go sideways as well as up or down.

How do you take this into consideration? How do the vertical and horizontal parts of his motion work together? And does gravity fit into the picture? These are the sorts of questions you will answer next.

Stephanie Skangos and DeAnna DelCarlo consider what other directions are possible for the diver upon release from a moving Ferris wheel.

High Noon

When the diver reaches the 12 o'clock position, his motion on the Ferris wheel is purely horizontal. If he were released at that moment, he would not be moving up or down at all, but only to the side.

Because of this, the diver's falling time would be the same as if he had fallen from a motionless Ferris wheel. But the diver will continue to move sideways throughout his fall. His sideways motion will be at the same rate as the platform was moving.

Use these facts, along with everything else you know about the Ferris wheel and the motion of falling objects, to answer these questions.

1. How long will it take the diver to reach the water level?

2. How far to the left of center will the diver be when he is 8 feet off the ground? In other words, what is his *x*-coordinate when he reaches the water level?

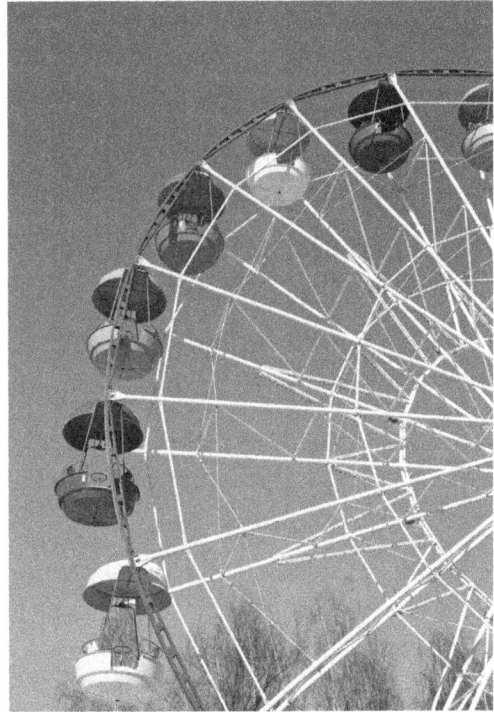

Leap of Faith

"Fire! Fire!" someone yells from down the hall. You reach for the doorknob, but remember to feel the door first. It's a good thing you did—because the door is extremely hot!

You are able to go out the window and make your way up to the roof, which is flat. As you contemplate your situation, the firefighters arrive.

"Jump! Jump!" they shout, and they hold out a rescue net. The net is 30 feet below you, and its center is 15 feet out from the edge of the roof.

You decide to run straight off the edge of the roof, hoping to go just far enough out to land in the middle of the net.

1. At what speed should you be traveling as you leave the roof in order to hit the center of the net? First figure out how long it will take you to fall the 30 feet.

2. If the net is 10 feet across, what range of speeds will allow you to hit the net?

The Ideal Skateboard

Let's return to the skateboard park from *A Simple Summary and a Complex Beginning.*

A skateboarder is holding on to the railing of a spinning merry-go-round platform and is moving in a circular path. Suddenly, the skateboarder lets go of the railing and heads toward a padded wall. Here are some other details about the situation:

- The platform has a 7-foot radius and makes a complete turn every 6 seconds.
- The skateboarder lets go from the 2 o'clock position.
- At the moment of release from the platform, the skateboarder is 30 feet from the padded wall.

1. How fast will the skateboarder travel? Assume, as before, that there is no friction to slow down the motion.

2. What is the angle shown in the diagram as θ?

3. How much closer will the skateboarder be to the wall after each second? In other words, what is the "toward the wall" component of the skateboarder's velocity?

4. Use your answer to Question 3 to find out how long it will take the skateboarder to reach the wall.

5. Find the actual distance the skateboarder travels. Use that information (and your answer to Question 1) as an alternate way to find out how long it will take the skateboarder to reach the wall.

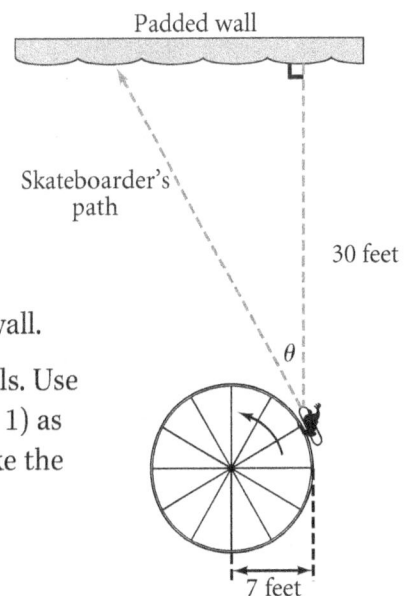

Padded wall

Skateboarder's path

30 feet

θ

7 feet

Racing the River

As part of their River Days festival, the citizens of River City are looking for a new event to raise money for charity. Someone suggests having groups of swimmers from the two local high schools, River High and New High, compete in a race across the river. The town would raise money by charging people to watch the race.

The river is generally quite calm when the festival is held, so the planners can safely assume that the current will not affect the swimmers. The race is set to take place along a straight stretch where the river is 200 meters across.

Representatives of New High point out that the River City swim team is based at River High, so most of the best swimmers attended that school. To even things out, the planners talk about making the River High swimmers swim farther. They discuss having the New High swimmers swim directly across the river while the River High swimmers swim at an angle of 45° off of the direct route, as shown here.

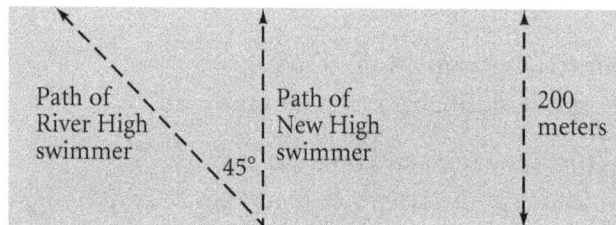

1. If a River High swimmer swims at a rate of 1.5 meters per second, how long will it take the swimmer to reach the other side?

2. How fast must a New High swimmer swim to reach the other side at the same time as a River High swimmer?

continued

The planners decide that the New High swimmers do not need such a big advantage. They change the path for the River High swimmers to make only a 30° angle with the direct route, as shown here.

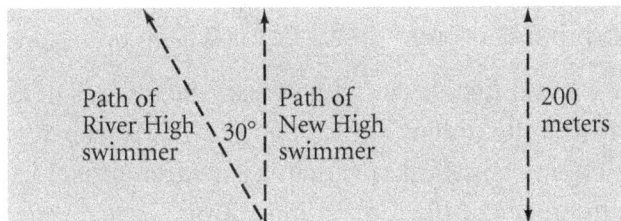

3. Answer these questions based on this new angle. Assume that a River High swimmer still swims at a rate of 1.5 meters per second.

   a. How far does a swimmer from River High need to swim to get across?

   b. How long will it take a River High swimmer to reach the other side?

   c. How fast must a New High swimmer swim to reach the other side at the same time as a River High swimmer?

One O'Clock Without Gravity

One night before the premiere of the show, our diver has a dream in which he is merrily spinning around on his Ferris wheel.

In this dream, the assistant decides to let go when the diver is at the 1 o'clock position. But in the dream, there is no gravity. So the diver sails up and to the left at a constant speed, in a direction tangent to the Ferris wheel's circumference at the 1 o'clock position.

Of course, his speed is the same as his speed when he was on the platform. As you know, that speed is 2.5π feet per second, or approximately 7.85 feet per second.

1. What is the vertical component of the diver's velocity? That is, how much height does he gain each second? As in other recent problems, you will probably need to find the angle labeled θ.

2. What is the horizontal component of his velocity? Remember, movement to the left is considered negative.

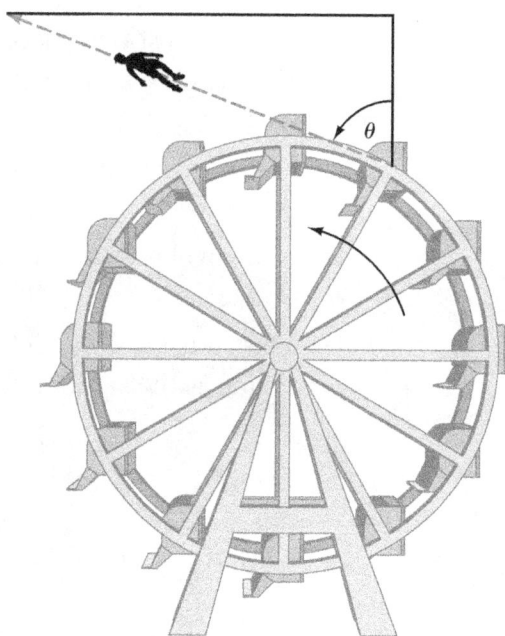

Swimming Pointers

Recall that the River High swimmer from Question 3 of *Racing the River* is swimming 1.5 meters per second in the direction 30° west of north. For now, ignore the distances (including the river's width) and focus on velocity.

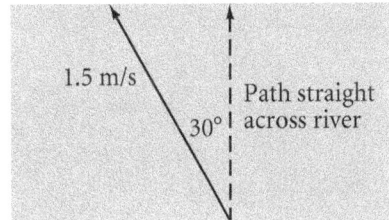

The diagonal arrow in the first diagram is a vector whose length (or **magnitude**) represents the swimmer's speed.

1. The vertical arrow in the second diagram is a vector representing the swimmer's rate of progress toward the opposite shore.

   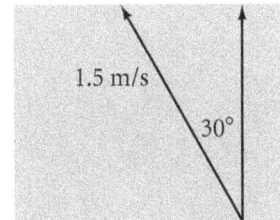

   a. How fast is the swimmer approaching the opposite shore?

   b. How is this quantity represented in the diagram?

2. a. What does the horizontal vector in the third diagram represent?

   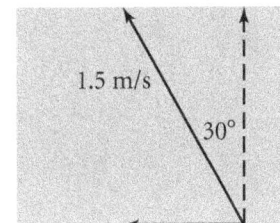

   b. What is that vector's magnitude?

3. Recall from *Complex Components* how to find the vector that represents the sum of two other vectors.

   a. Draw a diagram showing how this River High swimmer's actual speed and direction is the vector sum of the vertical and horizontal components of his velocity.

   b. Verify that these three vectors satisfy the Pythagorean theorem.

Vector Velocities

1. A car is traveling v miles per hour in the direction θ degrees north of east.

   a. Find the vertical and horizontal components of the car's velocity.

   b. Do these three velocities satisfy the Pythagorean theorem? Explain.

2. A certain Ferris wheel is turning counterclockwise. At a particular instant, a rider is moving horizontally to the right at 2.5 feet per second and vertically up at 6 feet per second.

   a. How fast is the rider actually moving?

   b. In what direction is the rider moving at that instant (to the nearest tenth of a degree)?

   c. What is the rider's clock position on the Ferris wheel at that instant (to the nearest minute)?

3. A swimmer is swimming across a river at s meters per second, headed directly toward the opposite shore. However, the river is flowing to the swimmer's right at r meters per second.

   a. Draw a vector diagram to describe the situation.

   b. To someone standing on a bridge above the river, how fast would the swimmer appear to be moving?

   c. In what direction is the swimmer actually moving toward the opposite shore?

Velocities on the Wheel

As you have seen, although the Ferris wheel's speed is constant, the vertical and horizontal components of the diver's velocity while still on the wheel are different for different positions in the cycle.

1. What are the horizontal and vertical components of the diver's initial velocity if he is released after 8 seconds on the Ferris wheel? (*Reminder:* The period for the Ferris wheel is 40 seconds. Also remember that for horizontal velocity, the positive direction is to the right.)

To generalize the situation, suppose the diver is released after W seconds.

2. First, assume that W is less than 10, so that the diver is still in the first quadrant when he is released. Write an expression in terms of W for the vertical and horizontal components of the diver's initial velocity.

3. Now consider all values of W from 0 to 40.

   a. For which values of W is the vertical component of velocity positive? For which values is it negative? For which values is it zero?

   b. For which values of W is the horizontal component of velocity positive? For which values is it negative? For which values is it zero?

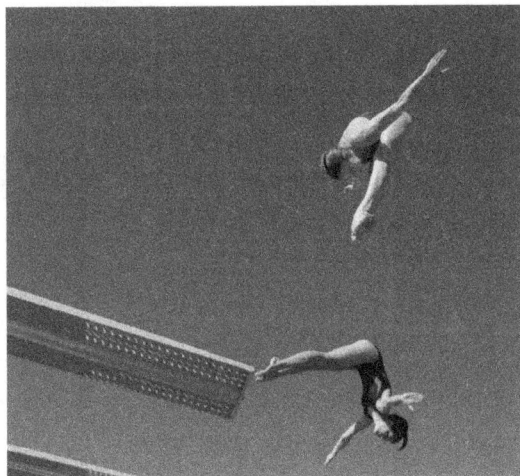

Release at Any Angle

In *Velocities on the Wheel,* you examined the
components of the diver's initial velocity if he is
released within the first quadrant. You may have
used a diagram like this one.

This analysis leads to these two equations:

- Vertical component of velocity = 7.85 cos (9*W*)
- Horizontal component of velocity = −7.85 sin (9*W*)

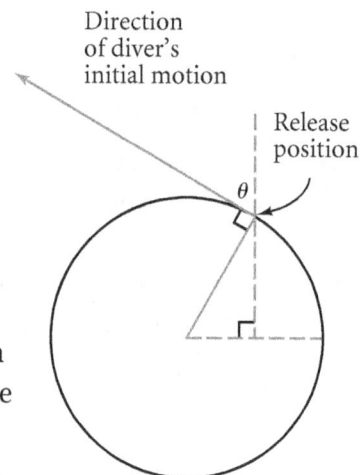

Direction
of diver's
initial motion

Release
position

Here, *W* is the time elapsed (in seconds) between when
the platform passes the 3 o'clock position and when the
diver is released, so the angle of turn is 9*W* degrees.

If the diver is still in the first quadrant, then *W* is less than
10 seconds. But what if *W* is more than 10 seconds? Do these
equations still work?

1. a. Assume for now that the equation for the vertical component
of the diver's initial velocity is correct for all values of *W*. Use
that equation to make a graph showing this vertical component
of velocity as a function of *W*, from *W* = 0 to *W* = 40.

   b. In Question 3a of *Velocities on the Wheel,* you examined how the
sign of the vertical component of velocity depends on *W*. Is your
graph consistent with your results from that question? Explain.

   c. For values of *W* between 10 and 20, the diver is released in
the second quadrant. Describe how the size of the vertical
component of velocity changes as *W* increases from 10 to 20.
Is your graph consistent with your description? Explain.

   d. Find the vertical component of velocity for
W = 11 and *W* = 12. Use diagrams as
needed to show how to get these values.
Are these values consistent with your
conclusions in part c? Explain.

2. Go through a sequence of steps for the
horizontal component of velocity similar to
those in Question 1.

An Expanded Portfolio of Formulas

As part of *A Simple Summary and a Complex Beginning,* you compiled a list of formulas that you used to solve the simplified version of the circus act problem. You also speculated on which of those formulas might need to be changed in the more complex version of the problem.

In your answers to these questions, label each formula appropriately and define each variable clearly so that you will know what each formula represents.

1. Reexamine your list of formulas from *A Simple Summary and a Complex Beginning.*

   a. Which of the formulas can you use as they are for the complex version of the problem?

   b. Which formulas need to be adjusted, and how do they need to be adjusted?

2. What completely new formulas do you need in order to solve the complex version of the problem?

Moving Diver at Two O'Clock

In *Three O'Clock Drop,* you were asked how long it will take the diver to reach the water level if he is released from the 3 o'clock position. In that activity, the diver's initial velocity had only a vertical component.

In *High Noon,* you answered the same question for a situation in which the initial velocity was all horizontal.

Now you will examine a Ferris wheel situation in which the diver's initial velocity is a blend of vertical and horizontal motion. Specifically, suppose the diver is released after a 30° turn on the Ferris wheel, at the moment when the platform reaches the 2 o'clock position.

1. How long will it take the diver to reach the water level?

2. What is the diver's *x*-coordinate when he reaches the water level? Remember that you are using a horizontal coordinate system in which the base of the Ferris wheel is zero and the positive direction is to the right.

3. What is the cart's *x*-coordinate when the diver reaches the water level?

The Danger of Simplification

In *Moving Cart, Turning Ferris Wheel,* you found that the assistant should release the diver after approximately 12.3 seconds. But that analysis was based on the idea that the diver would fall straight down as if released from a stationary platform.

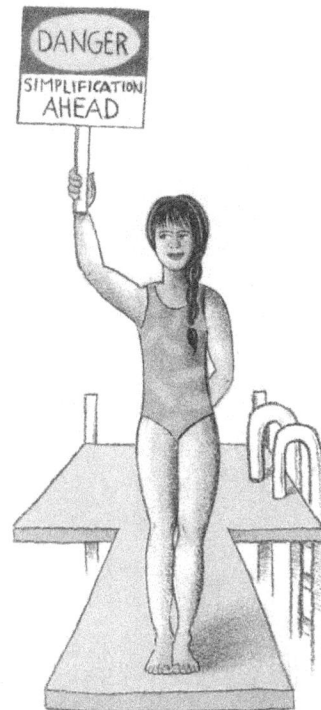

You now know that the Ferris wheel's motion will cause the diver to have an initial velocity of 2.5π feet per second (or about 7.85 feet per second) and that this initial velocity has both a vertical and a horizontal component. The initial velocity affects both the amount of time the diver is in the air and the diver's x-coordinate at the moment he reaches the water level.

To learn whether the initial velocity really matters, suppose the diver were released after 12.3 seconds (the time found in *Moving Cart, Turning Ferris Wheel*). Answer these questions, taking the initial velocity into account.

1. How long will it take the diver to fall to the level of the water in the cart?

2. What will the diver's x-coordinate be at the moment he reaches the water level?

3. What will the cart's x-coordinate be at the moment the diver reaches the water level?

4. Will the diver land in the tub of water?

The Diver Really Returns

At last! It's time to put all the formulas and ideas together and figure out exactly when the assistant should let go of the diver. The diver will certainly appreciate your hard work and careful analysis.

Ben Davis, Rashad Albert, Adé Thomas-Stewart, and Joaquin Menjivar-Austerlitz prepare their group presentation on the unit's final solution.

The Diver's Success

You solved a simpler version of the main unit problem in *Moving Cart, Turning Ferris Wheel.* Now you will take into account the fact that the diver leaves the Ferris wheel with an initial velocity that has both horizontal and vertical components.

Again, here are the details you need to know:

- The Ferris wheel has a radius of 50 feet.
- The center of the Ferris wheel is 65 feet above the ground.
- The Ferris wheel turns counterclockwise at a constant rate, making a complete turn every 40 seconds.
- When the cart starts moving, it is 240 feet to the left of the Ferris wheel's base.
- The cart moves to the right along the track at a constant speed of 15 feet per second.
- The water level in the cart is 8 feet above the ground.
- The cart starts moving as the platform passes the 3 o'clock position.

As before, let $t = 0$ represent the time when the Ferris wheel passes the 3 o'clock position, which is also when the cart begins moving. Let W represent the number of seconds that elapse between $t = 0$ and the moment the diver is released. Your task is to answer this question:

For what choice of W will the diver land in the tub of water?

Your group should prepare an oral report on your conclusions and how you reached them. You should also prepare your own write-up of your solution.

Feel free to use all the accumulated formulas and shortcuts you have developed. You may want to give variable names to complex expressions that are part of your solution. If you use a calculator to graph an equation, you can define some of these expressions as preliminary functions and then express your main equation or function in terms of these.

A Circus Reflection

You saw in *The Danger of Simplification* that if the diver's assistant uses the answer to the simplified problem, the result will be a serious error in the circus act.

1. Were you surprised at how big a difference the diver's velocity at the release makes in the outcome? Comment on your reaction to the result.

2. Describe another situation in which oversimplification of the mathematics of a problem might lead to serious consequences.

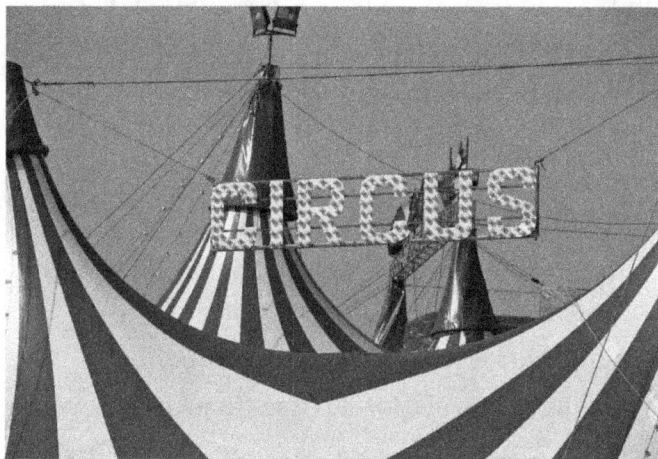

Beginning Portfolio Selection

Three of the big topics in this unit are somewhat related:

• Quadratic equations

• Complex numbers

• Vectors

Choose two or three activities that show some connection between any two of these topics. For each activity you choose, do these things:

• Describe in your own words what that connection is.

• Describe how the activity helped you make the connection.

The Diver Returns Portfolio

You will now put together your portfolio for *The Diver Returns*. This process has three steps:

* Write a cover letter that summarizes the unit.
* Choose papers to include from your work in the unit.
* Discuss your personal mathematical growth during the unit.

Cover Letter

Look back over *The Diver Returns* and describe the central problem of the unit and the key mathematical ideas. Your description should give an overview of how the key ideas—such as extending the sine and cosine functions and finding falling-time functions—were developed and how they were used to solve the central problem.

In compiling your portfolio, you will select some activities you think were important in developing the unit's key ideas. Your cover letter should include an explanation of why you selected each item.

continued

Selecting Papers

Your portfolio for *The Diver Returns* should contain these items:

- *Moving Cart, Turning Ferris Wheel*
- *An Expanded Portfolio of Formulas*
- *The Diver's Success*
- POW 1: *Which Weights Weigh What?*
- *Beginning Portfolio Selection*

Include in your portfolio the activities you discussed as part of this activity.

Personal Growth

Your cover letter for *The Diver Returns* should describe how the mathematical ideas develop in the unit. In addition, write about your own personal development during this unit. You may want to address this question:

> *What have you learned about solving a problem that is as complex as the circus act problem and that can't be solved by hand?*

Include any other thoughts about your experiences that you wish to share with a reader of your portfolio.

SUPPLEMENTAL ACTIVITIES

The supplemental activities for this unit follow up on your work with complex numbers. Here are some examples:

- *The Polar Complex* and *Polar Roots* develop and use the polar form of complex numbers.

- *Number Research* asks you to investigate different sets of numbers and how they relate to one another.

Complex Conjugation

Suppose $a + bi$ is a complex number, with a as its real part and bi as its imaginary part. (This is called the *rectangular form* of a complex number, in contrast to the *polar form*.) The *complex conjugate* (or, simply, the *conjugate*) of $a + bi$ is the complex number $a - bi$. For example, the conjugate of $2 + 3i$ is $2 - 3i$.

1. Find the sum and product of $a + bi$ and its conjugate.

2. Use your results to help you find a way to divide two complex numbers in rectangular form. That is, if $a + bi$ and $c + di$ are two complex numbers, what is the rectangular form of the quotient $(a + bi) \div (c + di)$?

3. In *Complex Numbers and Quadratic Equations*, you showed that $1 + i$ is a solution to the equation $x^2 - 2x + 2 = 0$. You then found the other solution. You also found both solutions to the equation $x^2 - 4x + 7 = 0$.

   a. Generalize these results using the idea of complex conjugates.

   b. Explain why your generalization is correct.

Absolutely Complex

For a complex number $a + bi$, we define its *absolute value* by the equation

$$|a + bi| = \sqrt{a^2 + b^2}$$

You will now investigate and prove some properties of the absolute value function.

1. Show that any complex number has the same absolute value as its conjugate. (The conjugate of $a + bi$ is $a - bi$.)

2. Suppose that u is a real number. Is the usual absolute value for real numbers, $|u|$, consistent with the previous definition of the absolute value for complex numbers? Explain.

3. Consider the equation $|uv| = |u| \cdot |v|$.

   a. Explain why this equation is true for all real numbers u and v. Illustrate your answer with examples.

   b. Is the equation $|uv| = |u| \cdot |v|$ also true for all complex numbers? Justify your answer and give examples.

4. Is it true for real numbers that $|u + v| = |u| + |v|$ What about for complex numbers? Explain your answers and illustrate with examples.

Niccolo Tartaglia (1500–1557) was an Italian mathematician and engineer. His work with cubic equations helped lead to the discovery of complex numbers.

The Polar Complex

In the Year 3 unit *High Dive,* you learned about **polar coordinates** as an alternative to **rectangular coordinates.** That same idea can be applied to complex numbers.

It is common to designate a complex number by the letter z. If the rectangular form of z is $a + bi$, its polar form is $r(\cos \theta + i \sin \theta)$, where $r \geq 0$.

1. Recalling what you know about polar representations, use this diagram to explain the relationship between the polar and rectangular forms of a complex number.

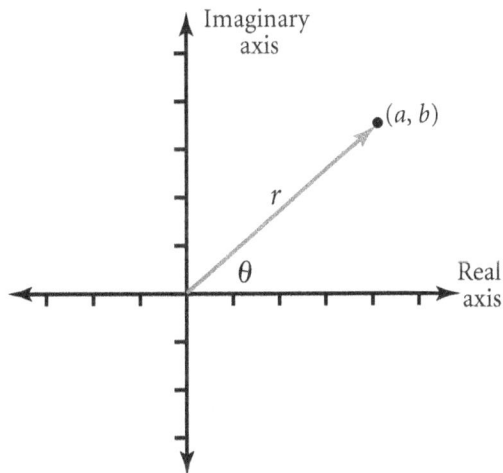

2. The equation $z = 2 + 5i$ is in rectangular form.

   a. Find a polar form for z.

   b. Graph both forms of z on the same set of axes. Is the result what you expected? Explain.

3. Let $z = 3(\cos 200° + i \sin 200°)$. Find the rectangular form of z.

4. What is the absolute value of $r(\cos \theta + i \sin \theta)$? Does this make sense? Explain.

continued ▸

One reason we use polar form is that it simplifies much of the arithmetic of complex numbers. Multiplying, dividing, and finding powers and roots are all a lot easier in polar form.

Suppose we designate the complex number z by its polar coordinates (r, θ), and do the same for z_1 and z_2. Then these simple equations apply:

$$z_1 \cdot z_2 = (r_1 \cdot r_2, \theta_1 + \theta_2) \qquad \left(\frac{z_1}{z_2} = \frac{r_1}{r_2}, \theta_1 - \theta_2\right)$$

If n is a positive integer, then

$$z^n = (r^n, n \cdot \theta)$$

5. Let $z = 1 - 2i$.

   a. Find a polar form for z.

   b. Use this polar form to find z^6.

Polar Roots

One of the most practical uses of the polar form of complex numbers is for finding roots. If the complex number z has a polar form (r, θ), then one of its nth roots is $\left(\sqrt[n]{r}, \frac{\theta}{n}\right)$.

For example, $z = -8 + 8\sqrt{3}i$ is in rectangular form. What is one of its fourth roots? First, change the equation into polar form:

$$r = \sqrt{(-8)^2 + (8\sqrt{3})^2} = \sqrt{64 + 192} = \sqrt{256} = 16$$

The angle θ is a second-quadrant angle whose cosine is $-\frac{1}{2}$, so $\theta = 120°$.

Then z has the polar form $(16, 120°)$ and has as one fourth root $\left(\sqrt[4]{16}, \frac{120°}{4}\right)$, or $(2, 30°)$.

1. Change $(2, 30°)$ into rectangular form. Verify that it is a fourth root of $-8 + 8\sqrt{3}i$.

A remarkable feature of numbers is that every complex number (which of course includes every real number) has exactly n distinct nth roots within the complex number system (assuming that n is an integer greater than 1). You might recall that a given point in the plane has many different polar representations. The same is true for complex numbers, and in the same way. It is this fact that enables us to find these n roots.

2. The expression $(16, 120°)$ is one of the polar forms of the complex number $z = -8 + 8\sqrt{3}i$ from Question 1. Find three other polar forms of z. Keep in mind that r must be nonnegative.

3. Use your results to find the other three fourth roots of z. If two of these turn out to be the same number (that is, to represent the same point in the plane), try some different polar forms of z.

4. Change at least one of the values you found in Question 3 into rectangular form, and verify that it is a fourth root of z.

5. Based on your findings, make a conjecture about the role of complex conjugates in nth roots of complex numbers.

Number Research

As you have seen in the mathematics courses you have taken, names are given to certain groups of numbers. For instance, the numbers 1, 2, 3, 4, 5, ... are sometimes referred to as the *counting numbers.* You have also worked with the set of *integers,* which are the numbers ... , −3, −2, −1, 0, 1, 2, 3, ...

In this unit, the activities *Imagine a Solution* and *Complex Numbers and Quadratic Equations* introduced other kinds of numbers.

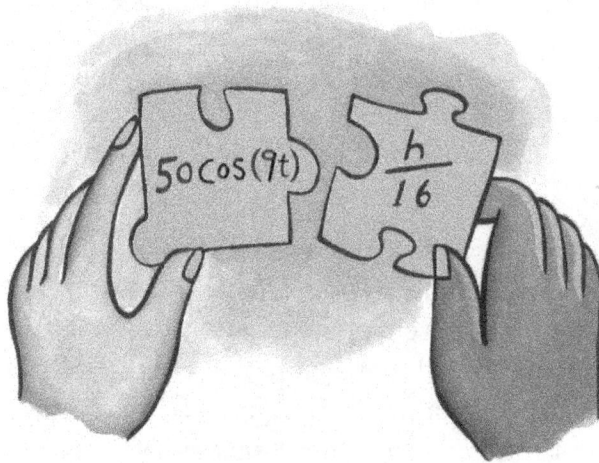

Your task here is to research names for different sets of numbers and learn about how those sets of numbers relate to one another. Then make a poster summarizing your findings.

Here are some additional names to get you started in your research:

- Whole numbers
- Rational numbers
- Imaginary numbers
- Complex numbers
- Transcendental numbers

PHOTOGRAPHIC CREDITS

Front Cover Photography

(upper row) Stephen Loewinsohn; (lower left and lower right) Shutterstock; (lower middle right and background image) iStockphoto

The Diver Returns

1 (upper row) Stephen Loewinsohn; (lower left and lower right) Shutterstock; (lower middle right and background image) iStockphoto; **3** Jerry Neidenbach; **8** Lynne Alper; **10** Lucidio Studio, Inc./SuperStock; **16** Stephen Loewinsohn; **18** Stephen Loewinsohn; **20** iStockphoto; **21** Shutterstock; **25** Shutterstock; **27** Stephen Loewinsohn; **31** Wikimedia Commons; **32** Peter Jonnard, Hillary Turner, Richard Wheeler; **33** Shutterstock; **34** Robert Brenner/PhotoEdit Inc.; **36** Shutterstock; **40** iStockphoto; **41** moodboard/Corbis; **42** PCN Photography/ Alamy; **46** Stephen Loewinsohn; **47** Stephen Loewinsohn; **48** iStockphoto; **54** Wikimedia Commons